TOO CUTE! BABY ANIMALS

BABY DESERT ANIMALS

RACHAEL MORLOCK

PowerKiDS press

Published in 2025 by The Rosen Publishing Group, Inc.
2544 Clinton Street, Buffalo, NY 14224

First Edition

Editor: Rachael Morlock
Book Design: Michael Flynn

Photo Credits: Cover Christian Mueller/Shutterstock.com, p. 1; (series background) Viktoriia LiSa/Shutterstock.com; p. 5 Kertu/Shutterstock.com; p. 7 Lubos K/Shutterstock.com; p. 9 Benny Marty/Shutterstock.com; p. 11 Elise Lefran/Shutterstock.com; p. 13 Stu Porter/Shutterstock.com; p. 15 atiger/Shutterstock.com; p. 17 Jason Yoder/Shutterstock.com; p. 19 seasoning_17/Shutterstock.com; p. 21 Gaston Piccinetti/Shutterstock.com.

Library of Congress Cataloging-in-Publication Data

Names: Morlock, Rachael, author.
Title: Baby desert animals / Rachael Morlock.
Description: [Buffalo, New York] : PowerKids Press, [2025] | Series: Too cute! Baby animals | Includes bibliographical references and index.
Identifiers: LCCN 2024003949 (print) | LCCN 2024003950 (ebook) | ISBN 9781499445176 (library binding) | ISBN 9781499445169 (paperback) | ISBN 9781499445183 (ebook)
Subjects: LCSH: Desert animals–Juvenile literature. | Desert animals–Infancy–Juvenile literature. | Desert animals–Habitat–Juvenile literature.
Classification: LCC QL116 .M67 2025 (print) | LCC QL116 (ebook) | DDC 591.754–dc23/eng/20240310
LC record available at https://lccn.loc.gov/2024003949
LC ebook record available at https://lccn.loc.gov/2024003950

Manufactured in the United States of America

CPSIA Compliance Information: Batch #CSPK25. For Further Information contact Rosen Publishing at 1-800-237-9932.

CONTENTS

Hot and Dry

A desert is a place that doesn't get much rain during the year. Few plants grow there, and only certain animals can **survive** in the hot and dry **climate**. From a young age, desert animals have what it takes to beat the heat!

Meerkat Pups

Meerkats live in Africa's Kalahari Desert. A mother meerkat gives birth to several pups in the rainy season. In the first few weeks, pups grow hair, open their eyes, and drink their mother's milk. While the mother feeds herself, other family members look after the pups!

Kangaroo Joeys

The red kangaroo needs very little water to survive in Australia's deserts. A baby kangaroo, called a joey, is the size of a cherry when it's born! The tiny joey crawls inside its mother's pouch. In the pouch, it can drink milk, stay safe, and grow.

Camel Calves

Camels can travel long **distances** without water in African and Asian deserts. A mother camel gives birth to one calf, usually during the rainy season. The calf can walk within half an hour of being born! Calves drink their mother's milk for about 18 months.

Caracal Kittens

The caracal survives in the desert by hunting in the cooler morning and evening hours. A **litter** of caracal kittens is born in a **burrow**. These kittens can't open their eyes for 10 days. They live on their mother's milk until she teaches them to hunt.

Fox Kits

Fennec foxes live in African and Arabian deserts. They sleep in cool underground dens during the hot daytime. Fennec kits are born in these dens and live there in family groups. Large ears, which give them great hearing, make this fox extra cute!

Owl Chicks

Great horned owls live in the Sonoran Desert of North America. They choose a high winter nest, often at the top of a saguaro cactus. A mother and father take turns keeping the eggs warm in the nest. Both guard and feed the chicks after they **hatch**.

Tortoise Hatchlings

The African spurred tortoise lives on the edge of the Sahara Desert. A mother tortoise digs several nests before choosing one. She lays 15 to 30 eggs and then covers them with sand or dirt. Tiny and **feisty** baby tortoises hatch eight months later.

Desert Babies

Desert animals have special skills for living in a hot and dry climate. Being **nocturnal** helps some deal with the heat. Many desert babies also count on their parents. Even in places that have little food and water, parents help their babies get what they need to grow.

GLOSSARY

burrow: A hole an animal digs in the ground for shelter.

climate: The average weather conditions of a place over a period of years.

distance: The space between two points.

feisty: Lively and not afraid to fight.

hatch: To break open or come out of an egg.

litter: A group of babies born to one animal at the same time.

nocturnal: Active mainly during the night.

survive: To keep living.

FOR MORE INFORMATION

BOOKS

Romero, Libby. *Desert Plants and Animals*. New York, NY: DK Publishing, 2023.

Weglinski, Michaela. *In the Desert*. Washington, D.C.: National Geographic Kids, 2020.

WEBSITES

Fennec Fox

www.nationalgeographic.com/animals/mammals/facts/fennec-fox

Learn about the world's smallest fox—the desert-dwelling fennec!

Great Horned Owl

www.audubon.org/field-guide/bird/great-horned-owl

Read this description of the great horned owl, from egg to adult.

Publisher's note to educators and parents: Our editors have carefully reviewed these websites to ensure that they are suitable for students. Many websites change frequently, however, and we cannot guarantee that a site's future contents will continue to meet our high standards of quality and educational value. Be advised that students should be closely supervised whenever they access the internet.

INDEX